Developing Num...

MEASURES, SHAPE AND SPACE

ACTIVITIES FOR THE DAILY MATHS LESSON

year

5

Dave Kirkby

A & C BLACK

Contents

Answers

Reprinted 2002, 2005

Published 2001 by A & C Black Publishers Limited

37 Soho Square, London W1D 3QZ

www.acblack.com

ISBN 0-7136-5882-7

Copyright text © Dave Kirkby, 2001

Copyright illustrations © Liz McIntosh, 2001

Copyright cover illustration © Charlotte Hard, 2001

Editors: Lynne Williamson and Marie Lister

The author and publishers would like to thank Madeleine Madden and Corinne McCrum for their advice in producing this series of books.

A CIP catalogue record for this book is available from the British Library.

A & C Black uses paper produced with elemental chlorine-free pulp, harvested from managed sustainable forests.

Printed in Great Britain by Caligraving Ltd, Thetford, Norfolk.

Introduction

Developing Numeracy: Measures, Shape and Space is a series of seven photocopiable activity books designed to be used during the daily maths lesson. They focus on the fourth strand of the National Numeracy Strategy *Framework for teaching mathematics*. The activities are intended to be used in the time allocated to pupil activities; they aim to reinforce the knowledge, understanding and skills taught during the main part of the lesson and to provide practice and consolidation of the objectives contained in the framework document.

Year 5 supports the teaching of mathematics by providing a series of activities which develop essential skills in measuring, and exploring pattern, shape and space. On the whole the activities are designed for children to work on independently, although this is not always possible and occasionally some children may need support.

Year 5 encourages children to:

- convert larger units of measure to smaller units;
- record estimates and readings from scales to a suitable degree of accuracy;
- understand area measured in square centimetres, and to understand and use the formula 'length x width' for the area of a rectangle;
- read the time on a 24-hour clock and use 24-hour clock notation, and to use timetables;
- recognise properties of 2-D shapes, and to make and visualise 3-D shapes;
- classify triangles;
- recognise reflective symmetry in polygons;
- recognise positions and directions, and to read and plot co-ordinates in the first quadrant;
- use a protractor to measure and draw angles to the nearest 5°.

Extension

Many of the activity sheets end with a challenge (**Now try this!**) which reinforces and extends the children's learning, and provides the teacher with the opportunity for assessment. On occasion, you may wish to read out the instructions and explain the activity before the children begin working on it. The children may need to record their answers on a separate piece of paper.

Organisation

Very little equipment is needed, but it will be useful to have available rulers, scissors, coloured pencils, counters, dice, squared paper, small clocks with moveable hands, geoboards, interlocking cubes, solid shapes, protractors and small mirrors. You will need to provide kitchen scales and a range of objects to be weighed for page 10, and straws (or sticks or paperclips) for page 35.

The children should also have access to measuring equipment to give them practical experience of length, mass and capacity.

To help teachers to select appropriate learning experiences for the children, the activities are grouped into sections within each book. However, the activities are not expected to be used in that order unless otherwise stated. The sheets are intended to support, rather than direct, the teacher's planning.

Some activities can be made easier or more challenging by masking and substituting some of the numbers. You may wish to re-use some pages by copying them onto card and laminating them, or by enlarging them onto A3 paper.

Teachers' notes

Very brief notes are provided at the foot of each page giving ideas and suggestions for maximising the effectiveness of the activity sheets. These can be masked before copying.

Structure of the daily maths lesson

The recommended structure of the daily maths lesson for Key Stage 2 is as follows:

Start to lesson, oral work, mental calculation	5–10 minutes
Main teaching and pupil activities *(the activities in the **Developing Numeracy** books are designed to be carried out in the time allocated to pupil activities)*	about 40 minutes
Plenary *(whole-class review and consolidation)*	about 10 minutes

4

Whole-class warm-up activities

The following activities provide some practical ideas which can be used to introduce or reinforce the main teaching part of the lesson.

Measures

Comparison activities

Ask the children to estimate the order of a set of objects based on a given measure. For example, provide a set of five different containers labelled A to E. The children, in pairs, decide which has the least capacity, the next least, and so on up to the most. The capacities are measured (in a chosen unit) and the true order determined. Compare this with the estimates.

Units activities

Call out measurements of length, mass or capacity, such as 450 cm, and ask the children to tell you an equivalent measurement using a different unit, for example: *4500 millimetres, 4·5 metres.*

Reading timetables

You need a simple train timetable (drawn on the board or on an OHT), and a changeable digital clock. Set the clock to a given time, for example 10:27. Show it to the children and say: *This is the time.* Ask a range of questions to practise reading the timetable: *How long do I have to wait for the next train? How long will it take me to reach Sheffield? If the next train is 20 minutes late, what time will I arrive at Chesterfield?*

Shape and space

'Yes/no' games

On the board, or on an OHT, draw a set of 2-D shapes in the cells of a 4 x 4 grid numbered 1–16. Think of a shape. The children should try to guess the shape by asking a series of questions to which you can only reply *yes* or *no*. Count how many questions are asked before the correct answer is given.

'Show me' shape activities

Give the children a sheet of paper showing a 4 x 4 grid with a different shape drawn in each cell. The children place this on their table and point to a shape according to instructions, for example: *Show me a hexagon; Show me a shape with eight vertices; Show me a shape with two pairs of parallel sides.*

'True or false' shape games

Give each child a pair of true and false cards. Make a true or false statement about the properties of a shape, for example: *A parallelogram has two pairs of parallel sides; An octahedron has six vertices; A kite is a type of quadrilateral; An isosceles triangle has more equal sides than does an equilateral triangle.* The children hold up a card to show whether the statement is true or false. Discuss the answers at each stage. As an alternative to cards, the children can show 'thumbs up' for true and 'thumbs down' for false.

Counter co-ordinates

Give each child a counter and a sheet of paper showing a co-ordinate grid (make the squares of the grid about two centimetres). The children use the counter to mark specified points on the grid, for example: *Show me the point which is 3 along and 4 up. Show me the point (5, 2). Show me a point which has a horizontal co-ordinate of 2 and a vertical co-ordinate of 7.*

Class co-ordinates

Seat the children in neat rows and columns, for example, 30 children in 6 rows and 5 columns. Give each column a horizontal co-ordinate (1 to 5), and each row a vertical co-ordinate (1 to 6). Play 'Stand up games', by giving instructions such as: *Stand up if your horizontal co-ordinate is 4. Stand up if your vertical co-ordinate is less than 3.* The children should sit down again between instructions.

Paper plate angles

Use two different-coloured circles of paper, identical in size (you could use two paper plates). Cut each circle with a straight line from the edge to the centre. Intersect the circles so that you can rotate one of them to demonstrate an angle. Show an angle to the children and ask them to estimate its size in degrees to the nearest 5°. Choose a child to measure the angle to the nearest 5° using a protractor. Award points for estimating within 5°, 10°, and so on. Repeat for different angles.

Clock angles

Give each child three cards labelled *acute*, *obtuse* and *reflex*. Choose a time, for example 9:50, and ask a child to show the time by turning the hands of an analogue clock. Now ask the children to hold up the card which describes the angle shown on the clock face. Repeat for different times.

Alien measures

• **Work with a partner. You will need a sheet each.**

☆ Estimate length ⓐ in **millimetres**.
☆ Measure the actual length with a ruler.
☆ Find the difference.
☆ Do the same for each length.
☆ Calculate the total difference to find your score.
☆ The winner is the player with the lowest score.

Record your answers on the chart.

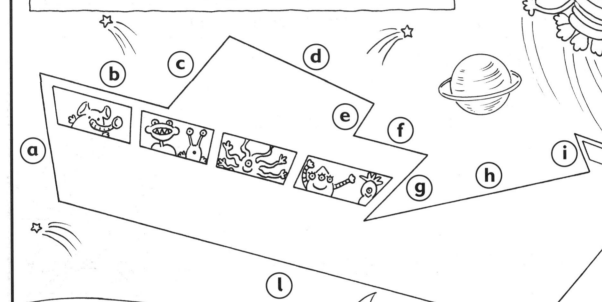

	a	b	c	d	e	f	g	h	i	j	k	l
Estimate (mm)												
Actual (mm)												
Difference (mm)												

• **Calculate the** | total difference | **to find your score.** _____

Now try this!

• **Write five lengths in** | millimetres | **.**

[ ] [ ] [ ] [ ] [ ]

• **Estimate each length by drawing a line.**

• **Measure the lines with a ruler to check.**

Teachers' note Check that the children understand how to play the game and know how to line up a ruler accurately, ensuring that the start of the ruler is exactly at the end of the line to be measured.

Developing Numeracy
Measures, Shape and Space
Year 5
© A & C Black

Insect Olympics

It is the long jump final for the insects! 100 cm = 1 m

- **Write these jumps in** centimetres .

Jump 1	1 m	100 cm
Jump 2	$\frac{1}{2}$ m	
Jump 3	$2\frac{1}{2}$ m	
Jump 4	2·3 m	

Jump 5	0·5 m	
Jump 6	$1\frac{1}{4}$ m	
Jump 7	$\frac{7}{10}$ m	
Jump 8	4 m	

- **Write these jumps in** millimetres . 1000 mm = 1 m

Jump 9	3 m	3000 mm
Jump 10	25 cm	
Jump 11	$4\frac{3}{4}$ m	
Jump 12	2·6 m	

Jump 13	$1\frac{1}{2}$ m	
Jump 14	0·9 cm	
Jump 15	8 cm	
Jump 16	$4\frac{1}{2}$ cm	

Now try this!

- **Round each jump to the** nearest metre .

For $\frac{1}{2}$ m, remember to round **up**.

Teachers' note Ensure that the children understand how to convert meters to centimetres (which requires multiplying by 100, thus sliding the digits two places to the left), and metres to millimetres (which requires multiplying by 1000, thus sliding the digits three places to the left).

Developing Numeracy
Measures, Shape and Space
Year 5
© A & C Black 2001

7

In the pool

The swimming pool measures 50 m x 20 m.

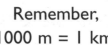
Remember, 1000 m = 1 km.

50 m

20 m

- **Write how many** metres **you travel if you swim:**

   1. 4 lengths _200 m_

   2. 5 widths _____

   3. 9 widths _____

   4. 15 lengths _____

   5. 6½ lengths _____

   6. 8½ widths _____

- **Write how many** lengths **you swim if you travel:**

   7. 100 m _____

   8. 1 km _____

   9. 650 m _____

   10. ½ km _____

- **Write how many** widths **you swim if you travel:**

   11. 500 m _____

   12. 2 km _____

   13. 1½ km _____

   14. 90 m _____

Now try this!

- **Write how many** metres **you swim in a week if you do:**

   8 lengths a day _____

   11 widths a day _____

Teachers' note Revise the fact that 1 km is the same as 1000 m.

Developing Numeracy
Measures, Shape and Space
Year 5
© A & C Black

- **Complete the statements.**

$1000 \text{ m} = 1 \text{ km}$

1. 1 km = _____ m

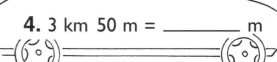

2. 5 km 300 m = _____ m

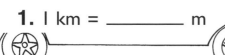

3. 2 km 800 m = _____ m

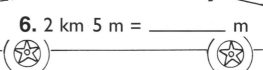

4. 3 km 50 m = _____ m

5. 4 km 190 m = _____ m

6. 2 km 5 m = _____ m

- **Write the distance in** ⟨metres⟩ **to each town.**

Town	Distance in metres
Tiptree	
Tenderfoot	
Hopeville	
Blackcap	
Denfield	
Tiffield	
Claw-on-Sea	
Hopton	

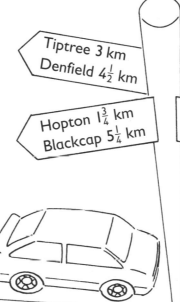

Tiptree 3 km
Denfield $4\frac{1}{2}$ km

Claw-on-Sea 5 km
Hopeville $3\frac{1}{4}$ km

Hopton $1\frac{3}{4}$ km
Blackcap $5\frac{1}{4}$ km

Tenderfoot 7·2 km
Tiffield 4·3 km

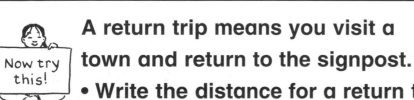

Now try this!

A return trip means you visit a town and return to the signpost.

- **Write the distance for a return trip to each town. Write in** ⟨metres⟩**, then in** ⟨kilometres⟩**.**

Teachers' note Ensure the children understand how to convert kilometres to metres (which requires multiplying by 1000, thus sliding the digits three places to the left).

Developing Numeracy
Measures, Shape and Space
Year 5
© A & C Black

Estimate the weight

• **Fill in the chart.**

You need:
kitchen scales
six objects to be weighed in grams

Object	My estimate (g)	It weighed (g)	Difference (g)
shoe	500 g	425 g	75 g

• **List your objects in order. Start with the lightest.**

• **Work out the total weight of all six objects. Then check on the scales.** _____ g

Now try this!

Teachers' note In order to help the children make estimates of the weight in grams, allow them to weigh another object accurately, then to compare the feel of the objects to be estimated with the object of known weight.

Developing Numeracy
Measures, Shape and Space
Year 5
© A & C Black

Cheesy eaters

- **Write the weight of each cheese in** $\boxed{\text{grams}}$.
- **Cut out the cards.**
- **Each mouse has eaten two cheeses. Match the cards.**

Remember, 1000 g = 1 kg.

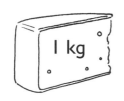

1 kg

_____

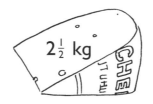

$2\frac{1}{2}$ kg

_____

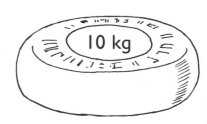

10 kg

_____

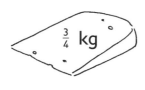

$\frac{3}{4}$ kg

_____

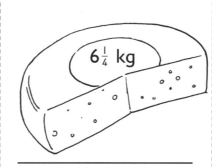

$6\frac{1}{4}$ kg

_____

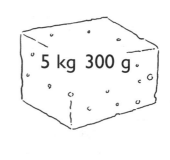

5 kg 300 g

_____

4·7 kg

_____

$3\frac{1}{4}$ kg

_____

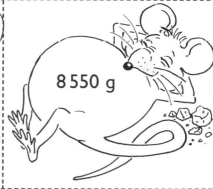

3 500 g

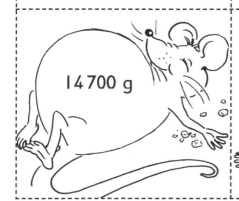

14 700 g

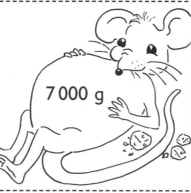

7 000 g

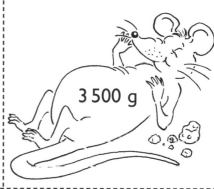

8 550 g

Teachers' note Remind the children that there are 1000 g in 1 kg, and discuss how they can use this to calculate the number of grams in fractions of a kilogram. As an extension activity, the children could be asked to find three cheeses which total 20 kg.

Developing Numeracy
Measures, Shape and Space
Year 5
© A & C Black 2001

Loopy scales

- **Cut out the cards.**
- **Put each answer next to the correct scales to make a loop.**

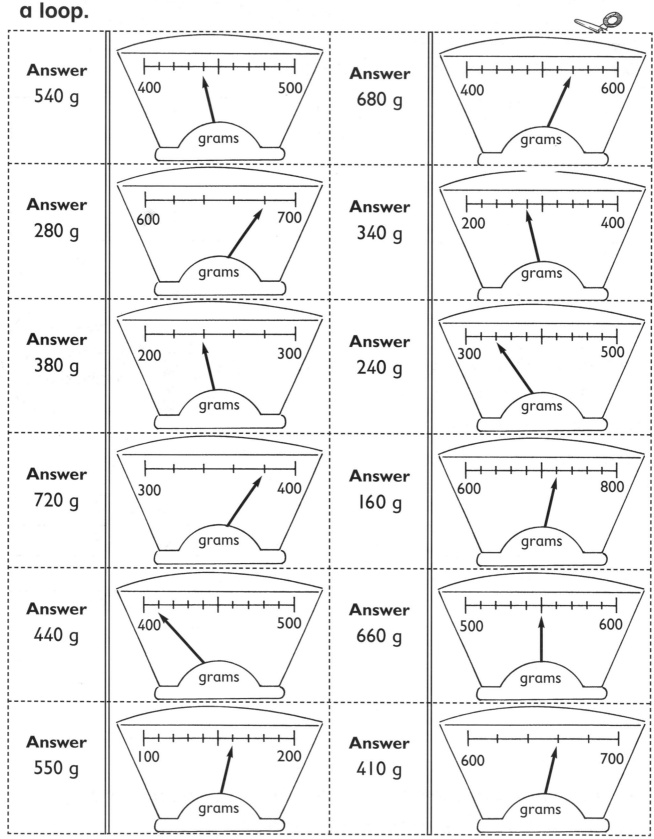

Teachers' note This could be an individual or group activity. Explain to the children that the cards form one continuous loop. Point out to the children that the scales are not all the same: some have 10 divisions separating 100 g, some have 5 divisions separating 100 g, and some have 10 divisions separating 200 g.

**Developing Numeracy
Measures, Shape and Space
Year 5**
© A & C Black

Estimating capacity

You need:
four different containers
a measuring cylinder
water or sand

- **Draw a picture of each container.**

a	**b**	**c**	**d**

- **Start with container** [a]. **Estimate the capacity in** [millilitres].

Record your answers on the chart.

- **Measure the capacity. Use a measuring cylinder and water or sand.**
- **Repeat for the other containers.**

	a	b	c	d
Estimate (ml)				
Actual (ml)				

- **Which was your best estimate?** ____
- **Which was your worst estimate?** ____

Now try this!

- **Draw two more containers.**
- **Estimate their capacities. Then measure them.**
- **Are your estimates improving?**

Teachers' note To help the children make reasonable estimates, show them a litre container and remind them that 1 litre is 1000 ml.

**Developing Numeracy
Measures, Shape and Space
Year 5
© A & C Black**

Measuring cylinders

• **Write each amount in** `millilitres` .

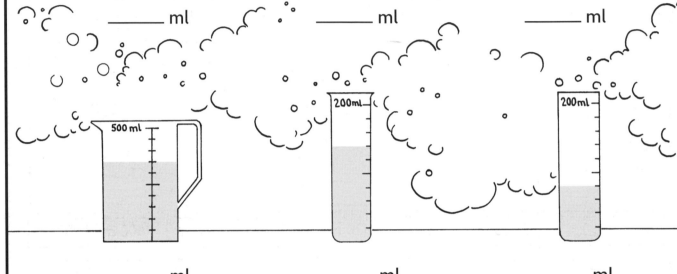

_____ ml _____ ml _____ ml

_____ ml _____ ml _____ ml

• **Colour each container to show the correct amount.**

1l	500ml	200ml
400 ml	150 ml	170 ml

Now try this!
• **Write the total amount in all nine containers.** _____
• **Write the answer in** `litres` **, as a decimal.** _____

Teachers' note Remind the children that they need to look closely at each cylinder, recognise its total capacity and identify the amounts represented by each of the marks.

Developing Numeracy
Measures, Shape and Space
Year 5
© A & C Black

Litres to millilitres

• **Colour the correct answer.**

1.
I litre =

100 ml	1000 ml	10 ml

2.
2½ l =

2000 ml	2500 ml	250 ml

3.
0·7 l =

70 ml	700 ml	7000 ml

4.
2¼ l =

2400 ml	2200 ml	2250 ml

5.
4 l 300 ml =

4300 ml	430 ml	4·3 ml

6.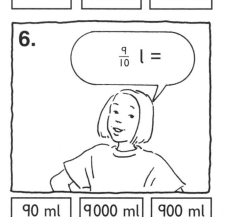
$\frac{9}{10}$ l =

90 ml	9000 ml	900 ml

7.
1·3 l =

130 ml	13000 ml	1300 ml

8.
1 l 50 ml =

150 ml	1·5 ml	1050 ml

9.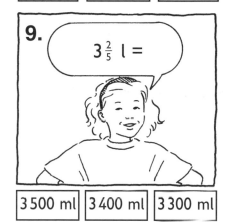
3⅖ l =

3500 ml	3400 ml	3300 ml

10.
4·6 l =

460 ml	4600 ml	46 ml

11.
3 l =

30 ml	300 ml	3000 ml

12.
4½ l =

4·2 ml	4002 ml	4500 ml

Teachers' note Ensure that the children know how to convert from litres to millilitres. The page could also be adapted to practise converting to centilitres instead of millilitres. Values could be altered to provide an easier or more challenging activity.

**Developing Numeracy
Measures, Shape and Space
Year 5**
© A & C Black

15

Thirsty family

I litre is **approximately** 2 pints.

- **Pour the water into the litre jug. Draw the level on the picture to the | nearest 50 ml |. Write the result.**

  I pint ≙ _____ ml

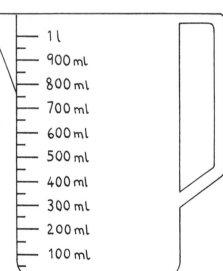

- **Use this to complete the chart. It shows how much the Quaff family drinks each day.**

	Pints	Litres
Baby Quaff	2 pints	
Mrs Quaff		2 litres
Mr Quaff		$4\frac{1}{2}$ litres
Quentin Quaff	3 pints	
Growler the dog	$\frac{1}{2}$ pint	

Now try this!

- **Find a large bucket. Estimate its capacity in | pints |.**
- **Check your estimate by pouring pints of water or sand into the bucket until it is full.**

Teachers' note Discuss that 1 litre is approximately 2 pints (more accurately $1\frac{3}{4}$ pints). Conversely, 1 pint is approximately $\frac{1}{2}$ litre (more accurately 570 ml). At the beginning of the activity, introduce the children to the 'approximately equals' sign. For the extension activity, the children will need a large bucket, a pint bottle, and water or sand.

Developing Numeracy
Measures, Shape and Space
Year 5
© A & C Black

Telly teasers

TV TODAY

Programme	Channel	Start time	Length
Kooky Bear	1	07:25	1 hour
Dinosaurs	4	11:50	25 mins
Bozz and Buzz	3	18:15	30 mins
Pet Choice	5	13:20	45 mins
Sport for Kids	34	11:45	20 mins
Comedy Half-hour	31	20:25	30 mins

• **Rewrite the start times using** am **and** pm .

1. Bozz and Buzz

6:15 pm

2. Sport for Kids

3. Kooky Bear

4. Comedy Half-hour

5. Pet Choice

6. Dinosaurs

Now try this!

• **Write the finish times using** 24-hour **time.**

Kooky Bear _____   Dinosaurs _____

Bozz and Buzz _____   Pet Choice _____

Sport for Kids _____   Comedy Half-hour _____

Teachers' note The children could be encouraged to show each time on an analogue clock with moveable hands.

Developing Numeracy
Measures, Shape and Space
Year 5
© A & C Black

Changing times

- **This game is for three players.**
- **Choose one of these time periods.**

☆ Cut out the cards. Spread them face down.
☆ Take turns to reveal a card.
☆ Say the new time using your chosen time period.
☆ Check with the other players. If you are right, keep the card. If not, replace it.
☆ The winner is the player with the most cards.

45 minutes later

25 minutes later

25 minutes earlier

 04:10

 05:15

 07:25

 15:30

 12:45

 21:10

 16:12

 09:46

 11:45

 08:50

 14:35

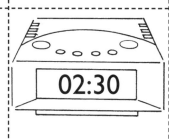 02:30

 20:15

 23:10

 16:17

 07:09

Teachers' note Each group of three children needs one copy of the sheet. If necessary, provide a list of answers for the children to check against. Children could also work on their own and time themselves. The game could be made easier or more challenging by changing the 'earlier' or 'later' times.

**Developing Numeracy
Measures, Shape and Space
Year 5**
© A & C Black

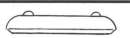

Train times

Liverpool to Durham	
Liverpool	0652
Warrington	0716
Birchwood	0722
Manchester	0745
Stalybridge	0801
Huddersfield	0821
Dewsbury	0832
Leeds	0847
York	0913
Thirsk	0934
Northallerton	0944
Durham	1019

Durham to Liverpool	
Durham	1837
Northallerton	1912
Thirsk	1923
York	1942
Leeds	2012
Dewsbury	2027
Huddersfield	2035
Stalybridge	2058
Manchester	2114
Birchwood	2134
Warrington	2140
Liverpool	2207

- **Look at the train timetables.**
- **Write how long these journeys take.**

1. Liverpool to Durham ___3 hrs 27 mins___

2. Durham to Liverpool _____

3. Manchester to Stalybridge _____

4. Leeds to York _____

5. Northallerton to Durham _____

6. Thirsk to Leeds _____

7. Birchwood to Liverpool _____

8. York to Durham _____

9. Dewsbury to Stalybridge _____

10. Durham to Leeds _____

 • **Draw a bar graph to show how long each stage of the journey from Liverpool to Durham takes.**

Teachers' note Discuss with the children that the convention for expressing 24-hour time varies, i.e. 23.15, 2315, 23:15. You could extend the activity by collecting real local timetables from the train or bus station. Ask the children to calculate journey times between pairs of stops.

**Developing Numeracy
Measures, Shape and Space
Year 5
© A & C Black**

Time tasks

• **Write how many** [seconds] **in:**

1. I minute _60 seconds_   2. 5 minutes _____
3. $\frac{1}{2}$ minute _____   4. I hour _____

• **Write how many** [minutes] **in:**

5. I hour _____   6. $3\frac{1}{2}$ hours _____
7. $2\frac{1}{4}$ hours _____   8. I day _____

• **Write how many** [hours] **in:**

9. I day _____   10. I week _____
11. a weekend _____   12. a morning _____

• **Write how many** [days] **in:**

13. I week _____   14. December _____
15. April _____   16. January to March _____

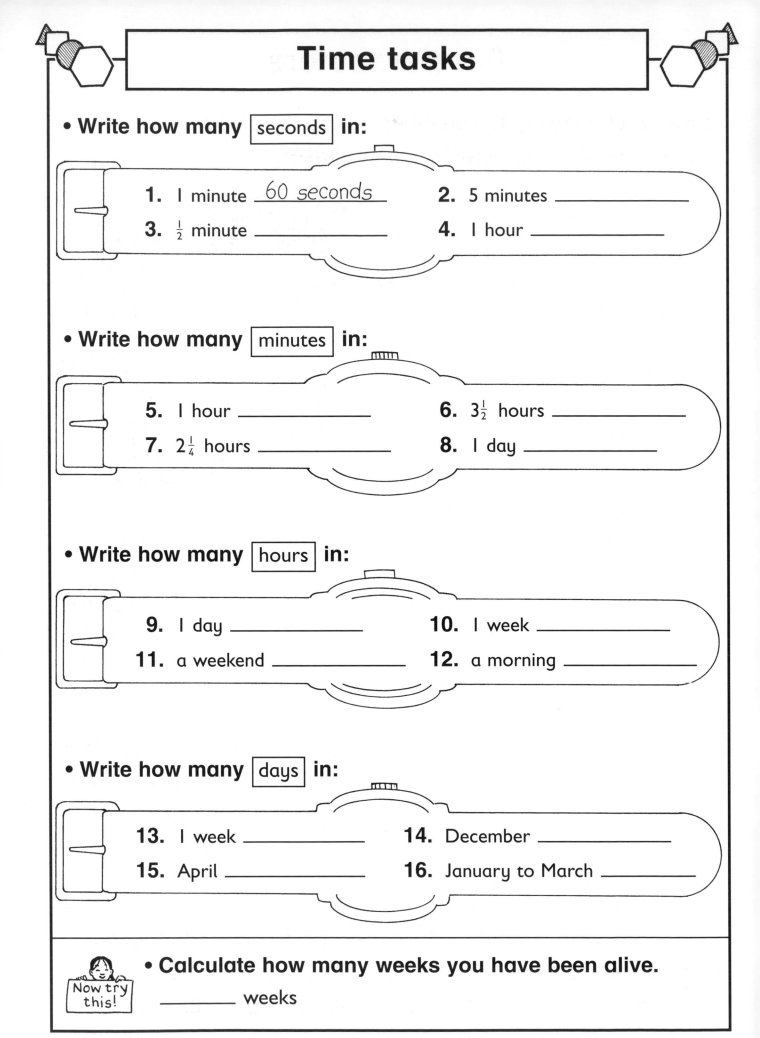

Now try this!

• **Calculate how many weeks you have been alive.**

_____ weeks

Teachers' note For the extension activity, children could find an approximate answer by using 1 year = 52 weeks. They could also use a calculator to work out how many days, hours or minutes they have been alive.

Developing Numeracy
Measures, Shape and Space
Year 5
© A & C Black

Pet perimeters

How much fencing do you need to stop the pets escaping?

• Calculate the ⎡perimeter⎤ of each garden.

1.
← 14 m →

10 m

48 m

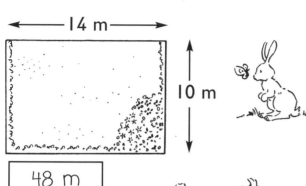

2.
← 20 m →

22 m

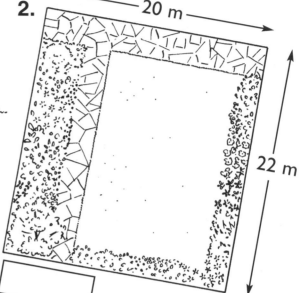

3.
← 9 m →

14 m

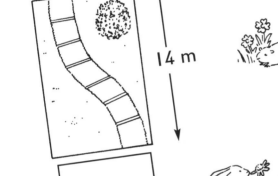

4.
← 25 m →

10 m

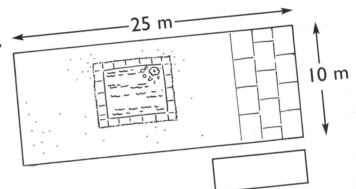

5.
← 15 m →

13 m

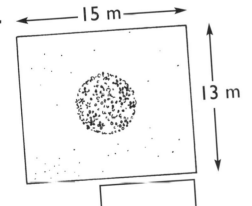

6.
← 16 m →

8 m

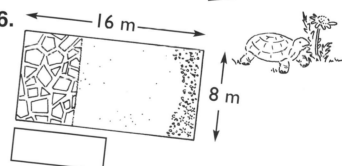

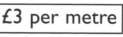
Teachers' note Explain to the children that their doubling skills will come in useful. To find the perimeter they need to double the length of the rectangle, then double the width, and total the two.

Developing Numeracy
Measures, Shape and Space
Year 5
© A & C Black

21

Triangle test

- **Estimate the** perimeter **of each triangle.**
- **Write them in order, shortest first.** C,_____
- **Measure each perimeter. Write your answer in** millimetres .

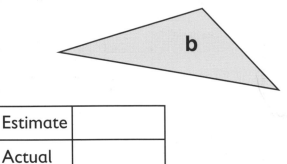

Estimate	140 mm
Actual	160 mm

Estimate	
Actual	

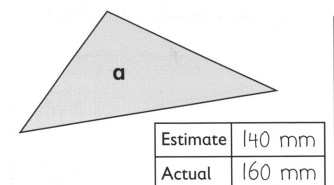

Estimate	
Actual	

Estimate	
Actual	

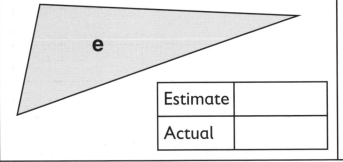

Estimate	
Actual	

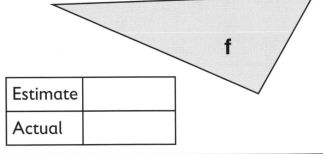

Estimate	
Actual	

- **Write the correct order.**

C,_____

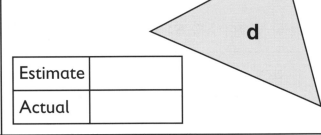

Were you close?

Now try this!

- **Draw three** quadrilaterals .
- **Ask a partner to estimate the perimeters, then measure them.**

Teachers' note Stress the need for accurate use of the ruler when measuring the sides of the triangles. The children could work in pairs, with a sheet each, to see whose estimate was closest. Note that some shapes have the same perimeter.

Developing Numeracy
Measures, Shape and Space
Year 5
© A & C Black

22

Area and perimeter

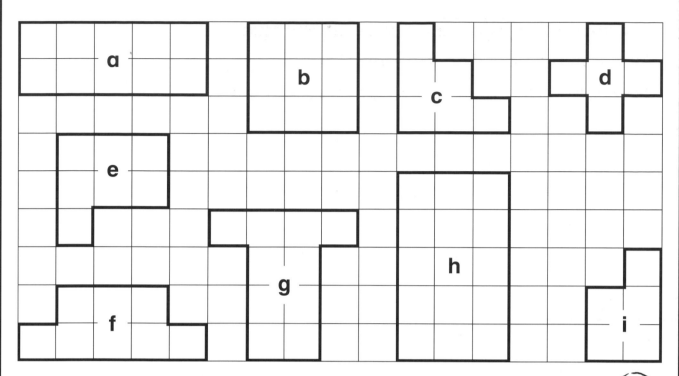

• **Look at the shapes. Complete the chart.**

Shape	Area	Perimeter	Number of sides	Name
a	10 cm²	14 cm	4	rectangle
	cm²	cm		
	cm²	cm		
	cm²	cm		
	cm²	cm		
	cm²	cm		
	cm²	cm		
	cm²	cm		
	cm²	cm		

Now try this!

• **Draw four different shapes which have an area of** 16 cm². **Use squared paper.**

• **Investigate whether they all have the same perimeter.**

Teachers' note Introduce the term dodecagon for a 12-sided shape. You could advise the children to mark the sides as they work out the perimeter so that they do not lose count. Point out that the numbers of sides are all even. Discuss whether it is possible to draw a shape on squared paper (with horizontal and vertical sides only) which has an odd number of sides, and why not?

Developing Numeracy
Measures, Shape and Space
Year 5
© A & C Black

Picture maths

This picture has been cut into eight pieces.

- Estimate the ⟨area⟩ of piece ⟨a⟩.
- Now measure it. Calculate the area.
- Repeat for each piece.

> Remember, area = width x length.
> 1 cm² → ⬜ 1 cm / 1 cm

	a	b	c	d	e	f	g	h
Estimate (cm²)								
Actual (cm²)								

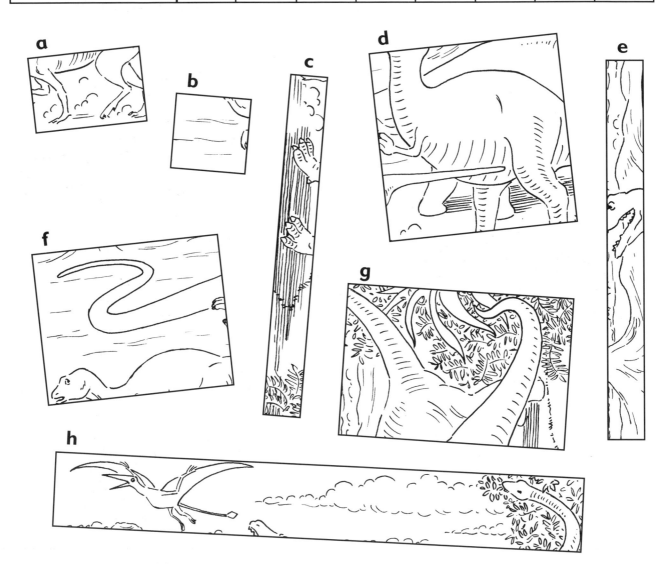

- Draw the rectangles onto squared paper to find the area of the whole picture. _____ cm²

Teachers' note Children are required to use the formula 'area of a rectangle = length x width' when measuring the area. One method of making a good estimate is to estimate the length and width in centimetres, then multiply them together. For the extension activity, some children may find it helpful to cut out the pieces to put together the whole picture.

Developing Numeracy
Measures, Shape and Space
Year 5
© A & C Black

Missing lengths

The ⬚ area ⬚ of each field is written inside it.

• Write the missing lengths.

1.

⬚ 9 m ⬚

45 m²   5 m

2.
4 m

36 m²   ⬚ m ⬚

3.
7 m

⬚ m ⬚

56 m²

4.
9 m

72 m²   ⬚ m ⬚

5.
9 m

81 m²

⬚ m ⬚

6.
⬚ m ⬚

60 m²   4 m

7.
⬚ m ⬚

160 m²   8 m

8.
4 m

48 m²   ⬚ m ⬚

Now try this!

• **Draw three different rectangles with a total perimeter of** ⬚ 36 cm ⬚ **.**
• **Find the area of each rectangle.**

What do you notice?

Teachers' note For more able children, the length/width of the fields could be masked. Only the area will be visible. The children can then estimate the length and width of the field.

Developing Numeracy
Measures, Shape and Space
Year 5
© A & C Black

Square challenge

The [perimeter] of this square is 28 cm.
The [area] of the square is 49 cm².

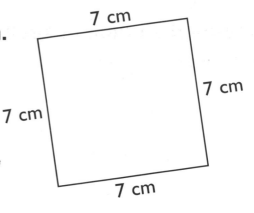

7 cm
7 cm
7 cm
7 cm

Remember,
area = width x length.

- **Write the** [area] **of squares with these perimeters.**

Perimeter	Area
8 cm	4 cm²
20 cm	
80 cm	
40 cm	
4 cm	
44 cm	

- **Write the** [perimeter] **of squares with these areas.**

Area	Perimeter
9 cm²	
64 cm²	
36 cm²	
144 cm²	
81 cm²	
900 cm²	

Now try this!

A rectangle has an area of [24 cm²].
- **What could its perimeter be?**

Find as many different perimeters as you can.

Teachers' note The children could write the area and perimeter of squares with sides 1 cm, 2 cm, 3 cm and so on, and then look for a pattern. Some children may need squared paper to help them.

Developing Numeracy
Measures, Shape and Space
Year 5
© A & C Black

Name the polygon

• **Write the name of each** polygon .

A polygon is a flat shape with straight sides.

Choose from the word-bank.

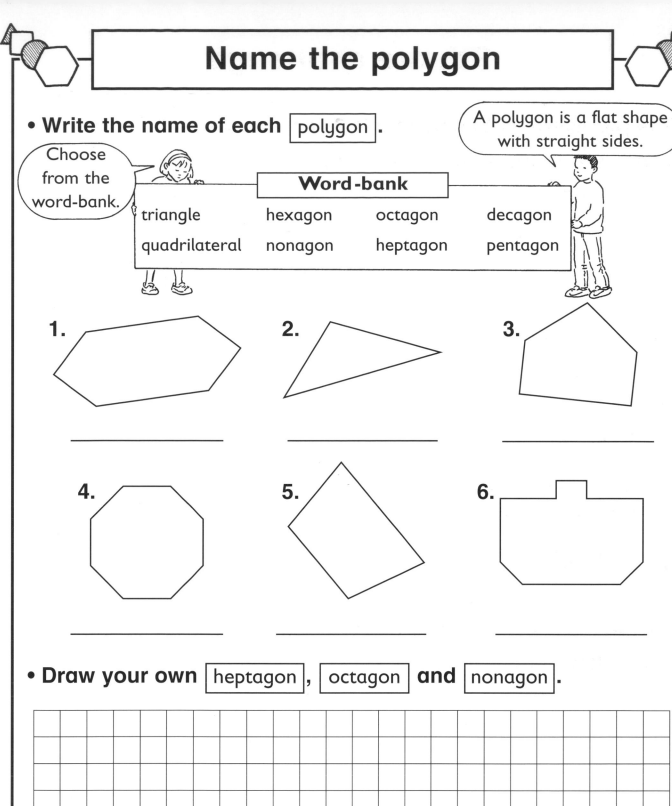

Word-bank

triangle   hexagon   octagon   decagon

quadrilateral   nonagon   heptagon   pentagon

1. _____

2. _____

3. _____

4. _____

5. _____

6. _____

• **Draw your own** heptagon , octagon **and** nonagon .

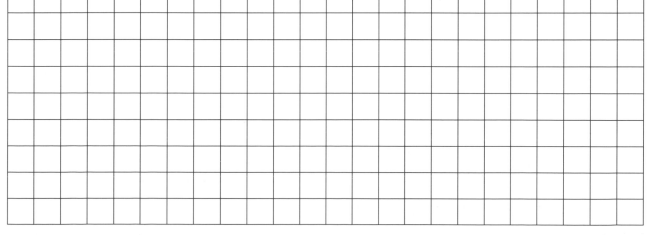

Now try this!

• **Draw a 12-sided polygon. This is called a** dodecagon .

Teachers' note Many 2-D shape resources are regular polygons. Children need plenty of experience in recognising irregular polygons, as in some of these examples. Encourage them to state whether each polygon is regular or irregular, and to explain the difference between regular and irregular shapes. If necessary, introduce or revise 'heptagon', 'nonagon' and 'decagon'.

Developing Numeracy Measures, Shape and Space Year 5 © A & C Black

Shape overlaps

- **Look at the squares. They overlap each other.**
- **Write down the shape of the overlapping part.**

1.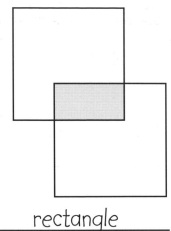

rectangle

2.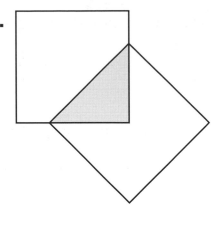

_____

3.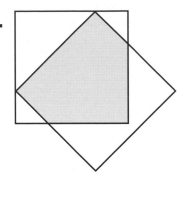

_____

4.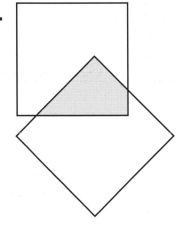

_____

5.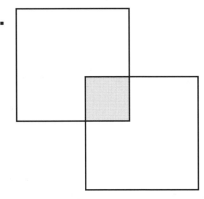

_____

6.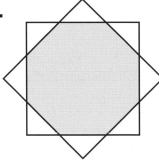

_____

Now try this!

- **Trace this square.**
- **Overlap the tracing with the original square to make:**

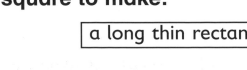
a long thin rectangle

a hexagon   a square

Teachers' note Encourage the children to state whether each shape is regular or irregular. As a
further extension activity, give the children tissue paper squares, identical in size but different in
colour. Ask them to overlap the squares and highlight the shape in the overlap. The most
effective way to view the overlap is on an overhead projector.

Developing Numeracy
Measures, Shape and Space
Year 5
© A & C Black

28

Change the sides

- **Join the dots to make different polygons with the given number of sides.**
- **Write the name of the polygon.**

Work in pencil first.

3 sides	4 sides	5 sides	6 sides	7 sides

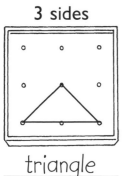 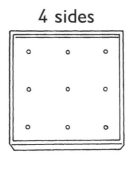 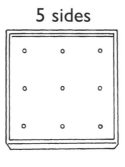 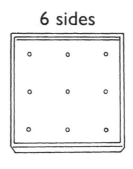 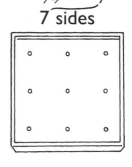

triangle _____ _____ _____ _____

3 sides	4 sides	5 sides	6 sides

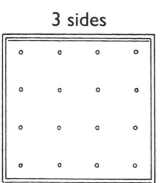 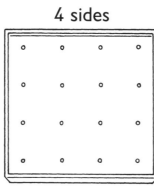 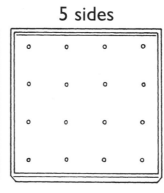 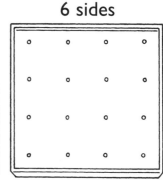

_____ _____ _____ _____

7 sides	8 sides	9 sides	10 sides

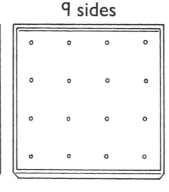

_____ _____ _____ _____

- **Investigate the** | maximum | **number of sides a polygon can have on a 4 x 4 dotty grid.**

Teachers' note Suggest to the children that they work in pencil first. Alternatively, they could experiment by creating shapes on a geoboard before drawing them on the grids. If necessary, introduce or revise 'maximum'.

**Developing Numeracy
Measures, Shape and Space
Year 5
© A & C Black**

Right-angled triangles

- **Ring the numbers of the** right-angled **triangles.**

  (1)   2   3   4   5   6   7   8   9   10

- **Mark the right angles.**

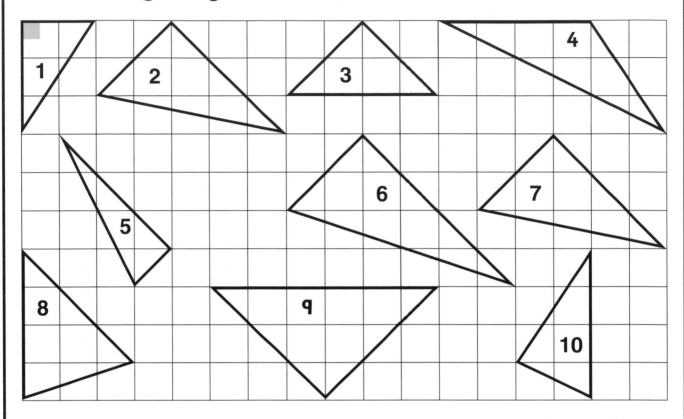

- **Draw three right-angled triangles. Each triangle has been started for you.**

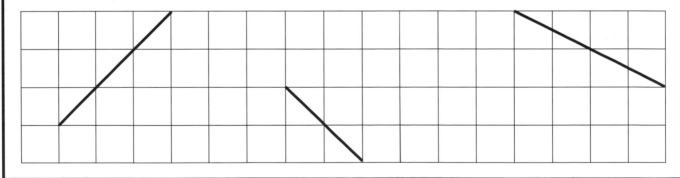

Now try this!
- **Draw three right-angled isosceles triangles. Make them all different.**
- **Mark the right angle.**
- **Draw the line of symmetry.**

Use squared paper.

Teachers' note Encourage the children to recognise and draw right-angled triangles whose right angles are formed by diagonal lines, rather than horizontal and vertical lines. Some children may need to measure the angles.

**Developing Numeracy
Measures, Shape and Space
Year 5**
© A & C Black

Types of triangle

- Complete the chart. Use these names to help you.

| scalene | right-angled | equilateral | isosceles |

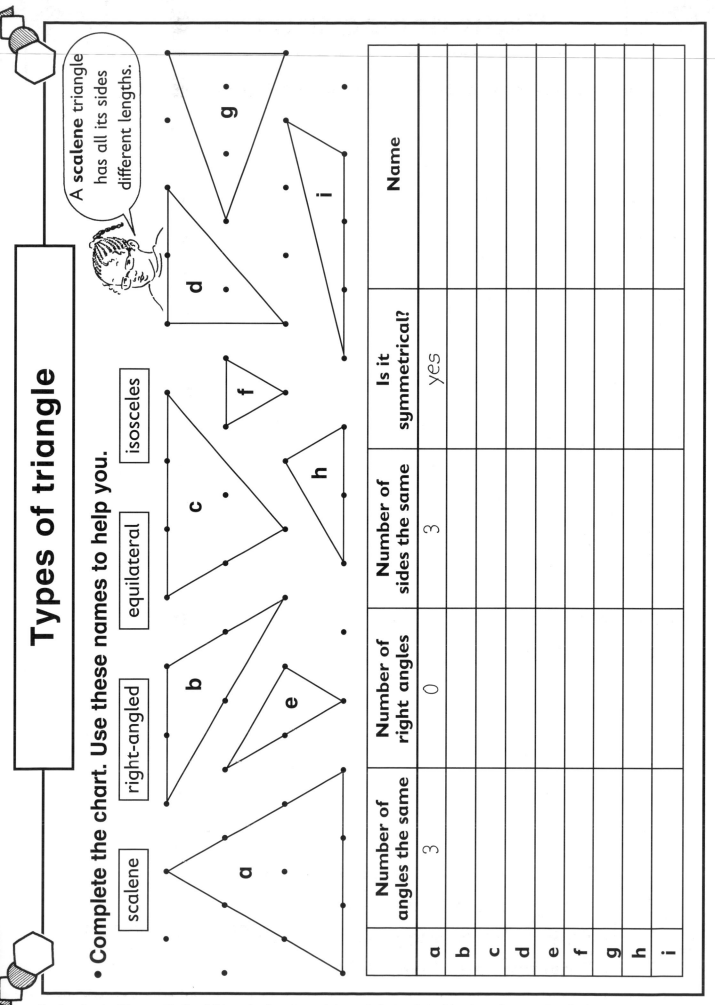

A **scalene** triangle has all its sides different lengths.

	Number of angles the same	Number of right angles	Number of sides the same	Is it symmetrical?	Name
a	3	0	3	yes	
b					
c					
d					
e					
f					
g					
h					
i					

Teachers' note Discuss with the children the fact that a right-angled triangle is either isosceles or scalene.

**Developing Numeracy
Measures, Shape and Space
Year 5**
© A & C Black

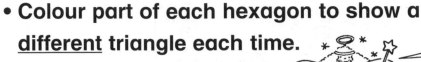

Tree decorations

- **Colour part of each hexagon to show a <u>different</u> triangle each time.**
- **Write the type of triangle underneath.**

> Try to make at least two of each type.

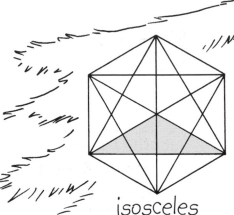

isosceles

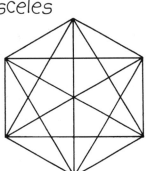

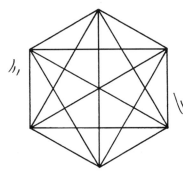

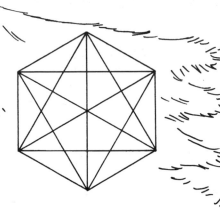

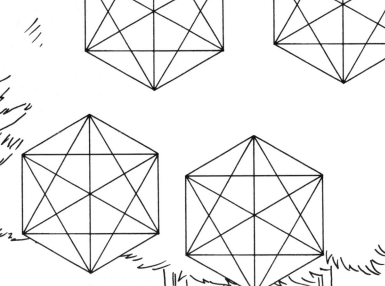

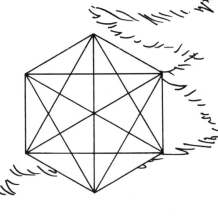

Now try this!

- **Draw four regular** pentagons **.**
- **Draw the diagonals, like this.**
- **Colour part of each pentagon to show a <u>different</u> triangle each time.**

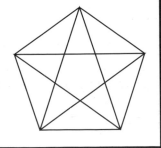

Teachers' note Emphasise the need for the triangles to be different. Children often produce two triangles which they think are different, but which are in fact the same triangle in a different position. Give examples if necessary. For the extension activity, provide regular pentagon templates.

Developing Numeracy Measures, Shape and Space Year 5
© A & C Black

Straw triangles

 **You need** short straws all the same length.

With seven straws you can make these triangles.

You **cannot** complete this one.

sides 1, 3, 3   sides 2, 2, 3

- Make triangles using the numbers of straws on the chart.

Each time, make as many **different** triangles as you can.

- Complete the chart.

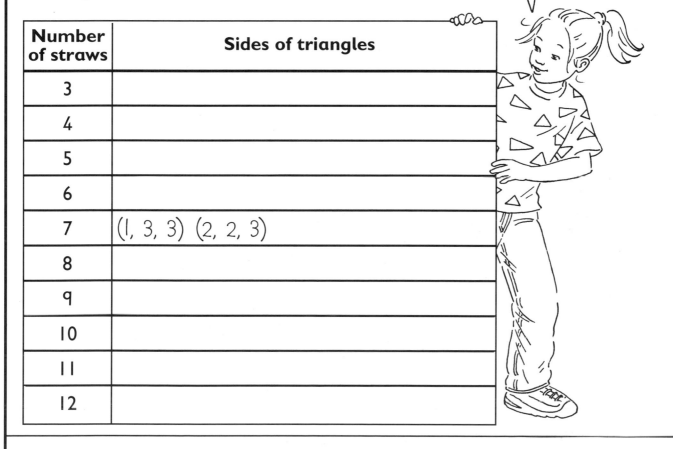

Number of straws	Sides of triangles
3	
4	
5	
6	
7	(1, 3, 3) (2, 2, 3)
8	
9	
10	
11	
12	

 **Now try this!**

- **Look at your completed chart. Ring:**

  isosceles triangles red   **equilateral** triangles blue

  scalene triangles orange

Teachers' note Provide the children with sets of identical short straws, sticks or paperclips. Emphasise the need for the triangles to be different each time. The activity can be extended to searching for quadrilaterals.

Developing Numeracy Measures, Shape and Space Year 5 © A & C Black

Dot to dot

- **Join the dots to make triangles.**

- **Draw six different** right-angled **triangles.**

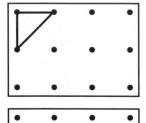

- **Draw six different** isosceles **triangles.**

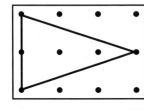

- **Draw six different** scalene **triangles.**

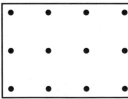

- **Mark all the** obtuse **angles like this.**
- **How many are there?** _____

36

Teachers' note Suggest to the children that they work in pencil first. Alternatively, they could experiment by creating shapes on a geoboard before drawing them on the grids. Emphasise the need for the triangles to be different. For the extension activity, define 'obtuse' if necessary (an angle more than 90° but less than 180°).

Developing Numeracy
Measures, Shape and Space
Year 5
© A & C Black

Shape guessing game

• **Play this game with a partner.**

☆ Choose one of the shapes below.
☆ Your partner has to guess which shape you have chosen by asking you questions. You can only answer 'yes' or 'no'.
☆ Record how many questions it takes to guess the shape.
☆ Play again, but this time your partner chooses.

Keep your shape a secret!

Teachers' note Each pair of children needs one copy of the sheet. The game is designed to encourage the use of shape vocabulary. Discuss words that will be useful, for example: side, vertex, right angle, parallel, perpendicular, triangle, quadrilateral, pentagon, symmetry. If appropriate, the children could also ask questions relating to the area or perimeter of the shapes.

**Developing Numeracy
Measures, Shape and Space
Year 5
© A & C Black**

37

Parallel sides

Li has drawn these shapes on squared paper.

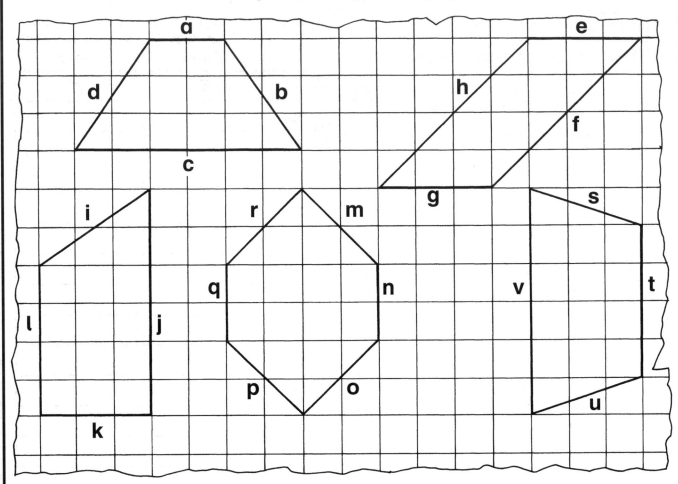

> **Parallel** sides are always the same distance apart.

- **Write which sides of the shapes are** | parallel |.

1. _a_ is parallel to **c**   2. ___ is parallel to **h**

3. ___ is parallel to **q**   4. ___ is parallel to **v**

5. ___ is parallel to **l**   6. ___ is parallel to **g**

7. ___ is parallel to **o**   8. ___ is parallel to **m**

9. ___ is parallel to **a**   10. ___ is parallel to **j**

Now try this!

- **Draw three shapes of your own which have parallel sides. Label the parallel sides.**

Teachers' note Discuss with the children the fact that shapes can have one pair of parallel sides (trapezium), two pairs of parallel sides (parallelogram), three pairs of parallel sides (regular hexagon), and so on.

Developing Numeracy
Measures, Shape and Space
Year 5
© A & C Black

Capital letters

- **Which capital letters have** $\boxed{\text{parallel}}$ **lines? Which have** $\boxed{\text{perpendicular}}$ **lines?**

- **Tick or cross the chart.**

> **Perpendicular** lines are straight lines at right angles to each other.

A B C D

E F G H

I J K L M

N O P Q R

S T U V

W X Y Z

	Parallel lines	Perpendicular lines
A	✗	✗
B		
C		
D		
E		
F		
G		
H		
I		
J		
K		
L		
M		
N		
O		
P		
Q		
R		
S		
T		
U		
V		
W		
X		
Y		
Z		

Now try this!

- **Write words using the letters you have ticked.**

Example: WET

Teachers' note Some letters can be argued for and against having perpendicular lines, for example K and X, depending on how they are written.

**Developing Numeracy
Measures, Shape and Space
Year 5**
© A & C Black

39

Paper chains

Make each pattern different.

- Draw all the diagonals of each polygon.
- Colour parts of each polygon to create a pattern.

- Draw round the outline of a different [regular] polygon.
- Draw all its diagonals. Colour parts to create a pattern.

Teachers' note Remind the children that to draw all the diagonals of a polygon, they should draw a straight line from each vertex to every other vertex.

40

Developing Numeracy
Measures, Shape and Space
Year 5
© A & C Black

Crazy paving

- **Draw all the diagonals of each polygon.**

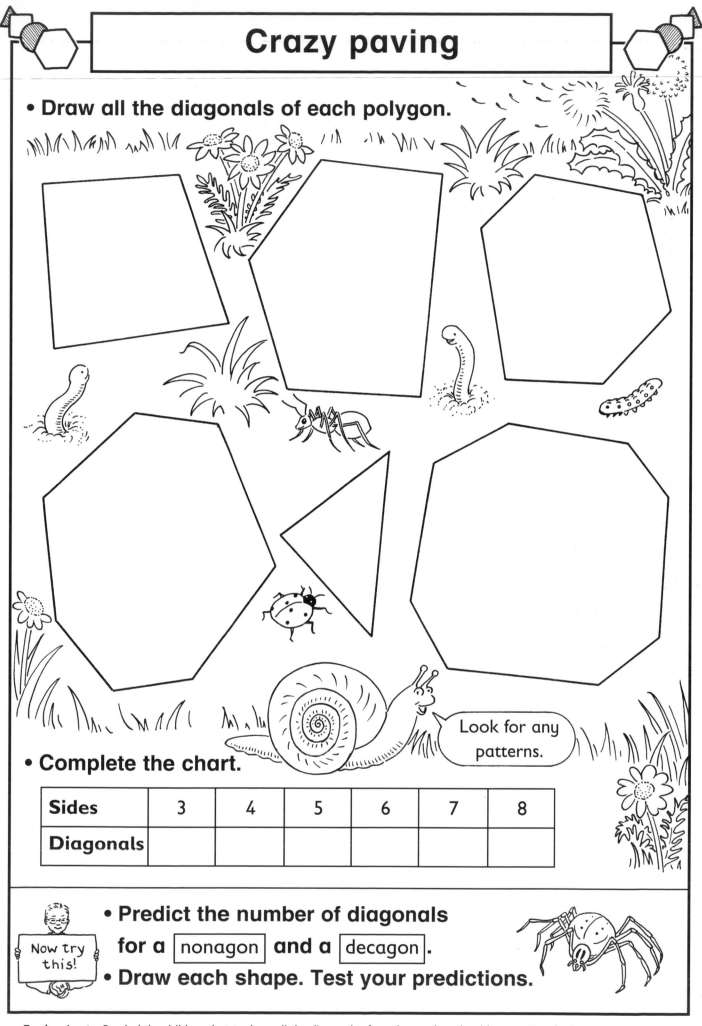

Look for any patterns.

- **Complete the chart.**

Sides	3	4	5	6	7	8
Diagonals						

Now try this!
- **Predict the number of diagonals for a** nonagon **and a** decagon .
- **Draw each shape. Test your predictions.**

Teachers' note Remind the children that to draw all the diagonals of a polygon, they should draw a straight line from each vertex to every other vertex. The children might find it useful to number the diagonals or keep a tally as they draw them. For the extension activity, revise 'nonagon' and 'decagon' if necessary.

Developing Numeracy
Measures, Shape and Space
Year 5
© A & C Black

41

Diagonals of a rectangle

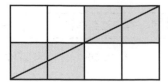

On a 2 x 4 rectangle the diagonal passes through **four** squares.

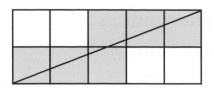

On a 2 x 5 rectangle the diagonal passes through **six** squares.

- **Draw other rectangles.**
- **Find out how many squares the diagonal passes through.**
- **Complete the chart.**

Draw the rectangles on this grid.

		Length of rectangle					
		1	2	3	4	5	6
Width of rectangle	1						
	2				4	6	
	3						
	4		4				
	5		6				
	6						

Now try this!

- **Write about any patterns you can see in the chart.**

Teachers' note The centre of the diagonal either passes through the vertices of four squares, or through the sides of two squares. Suggest to the children that they draw the rectangles on the grid in pencil so that they can be rubbed out when more space is needed. Alternatively, provide extra sheets of squared paper.

Developing Numeracy
Measures, Shape and Space
Year 5
© A & C Black

Name the shape

- **Write the name of each solid shape.**

Choose from the word-bank.

Word-bank

cylinder	cuboid	cube
pyramid	prism	cone

1.

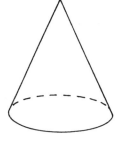

_____

2.

_____

3.

_____

4.

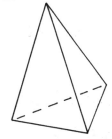

_____

5.

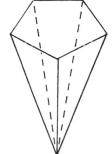

_____

6.

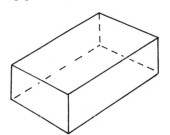

_____

7.

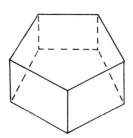

_____

8.

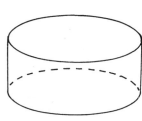

_____

9.

_____

10.

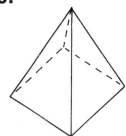

_____

11.

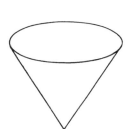

_____

12.

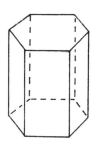

_____

Now try this!

- **For each prism and pyramid, write which type it is.**

Example: pentagonal prism   square-based pyramid

Teachers' note A cube is a special case of a cuboid. Similarly, a cuboid is a rectangular prism.

Developing Numeracy
Measures, Shape and Space
Year 5
© A & C Black

• **Follow the instructions.**

☆ Cut out each net. Fold it along the lines.

☆ Find the nets that fold to make an **open** cube. Colour red the square that makes the base of the cube.

☆ Find the net that folds to make a **closed** cube. Colour red the squares that make the top and the base of the cube.

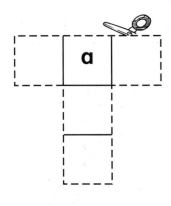

a

b

c

d

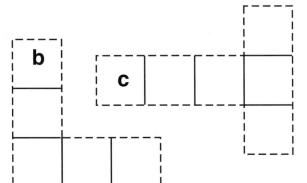

e

f

g

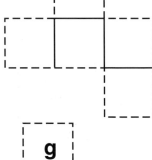

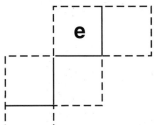

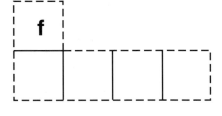

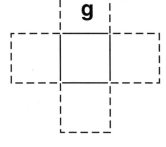

h

i

m

j

k

l

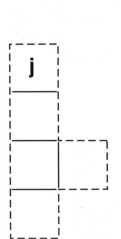

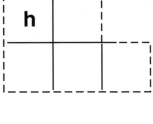

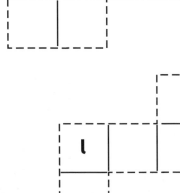

Teachers' note If possible, enlarge the sheet to A3 size to make the nets easier to handle. As an extension, the children could explore other nets for closed cubes (using six squares).

**Developing Numeracy
Measures, Shape and Space
Year 5
© A & C Black**

Cuboid models

- **Use interlocking cubes to build model** a.
- **Write on the chart how many cubes you needed.**
- **Now write the fewest cubes you need to add to make it into a cuboid.**

> Do the same for each model.

	a	b	c	d	e	f	g	h	i
Number of cubes	4								
Number of cubes to make a cuboid	4								

a

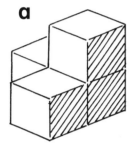

b

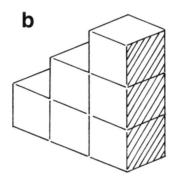

c

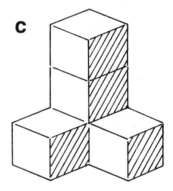

d

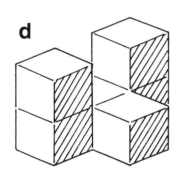

e

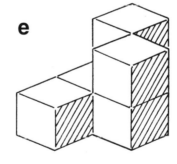

f

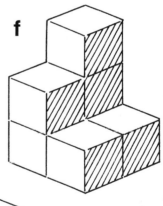

g

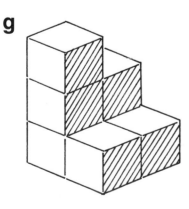

h

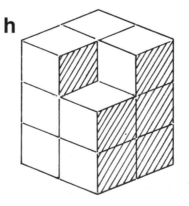

i

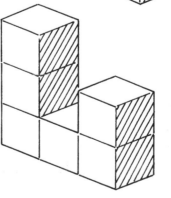

Teachers' note Once the children have built the models, they can try to calculate the surface area of each model.

**Developing Numeracy
Measures, Shape and Space
Year 5
© A & C Black**

45

Plaiting a tetrahedron

- **Follow these instructions to make a** tetrahedron .

☆ Cut out the net. Fold it along the lines.
☆ Begin plaiting: place 1 over 2
                  place 3 under 4
                  fold the rest around
                  tuck 5 underneath 1.

- **Now unfold it. Write
a different property of
a tetrahedron on each
blank face.**

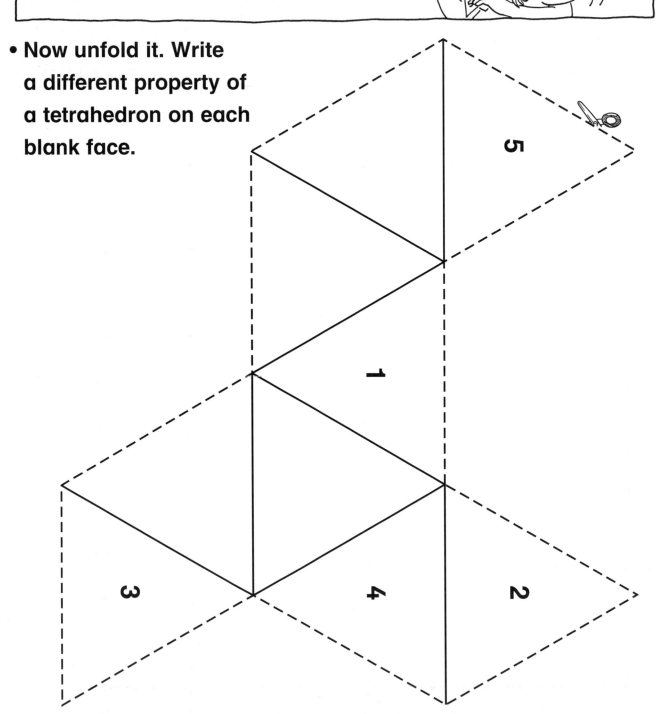

Teachers' note This tetrahedron is regular, and it is a type of triangular-based pyramid. As an extension activity, the children could design a different net for a tetrahedron.

46

**Developing Numeracy
Measures, Shape and Space
Year 5
© A & C Black**

Trying for triangles game

● **Play this game with a partner.**

☆ Take turns to throw a dice twice. This gives you a **horizontal** and a **vertical** co-ordinate.

Example: 1st throw   2nd throw

  (4, 2)

☆ Mark the co-ordinate with a counter on your grid.
☆ The winner is the first player to make an isosceles triangle **and** a right-angled triangle with their counters.

You need:
a copy of the grid each
a dice
20 counters each

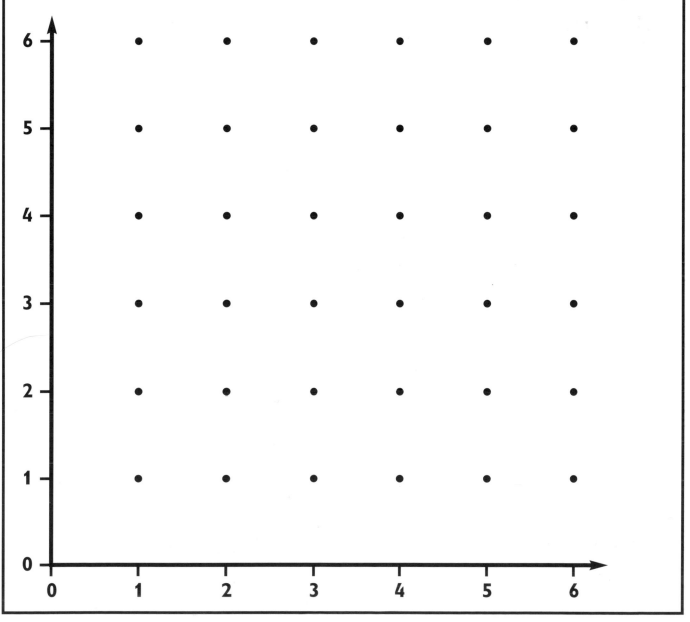

Teachers' note The children need a copy of the sheet each. Help them to remember that the horizontal co-ordinate comes first by using the phrase 'along the corridor, then up the stairs'. This game could also be adapted to find squares, rectangles and so on.

Developing Numeracy
Measures, Shape and Space
Year 5
© A & C Black

Treasure island

• **Look at the map.**

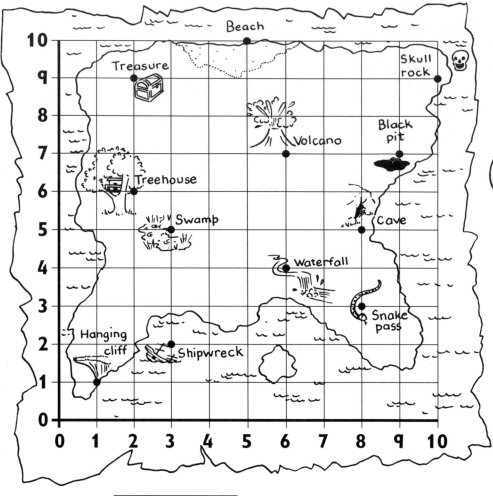

Read across the grid and then up.

• **Write the** co-ordinates **of these points.**

1. Shipwreck ___(3, 2)___
2. Swamp _____
3. Snake pass _____
4. Hanging cliff _____
5. Treehouse _____
6. Volcano _____
7. Skull rock _____
8. Waterfall _____
9. Treasure _____
10. Cave _____
11. Beach _____
12. Black pit _____

Now try this!

• **Write** co-ordinates **for these points.**

Boat _____   Hut _____

Hill _____   Wood _____

• **Plot them on the map.**

Teachers' note The children may need reminding of how to plot co-ordinates. They could draw their own treasure maps on squared paper and mark on specific co-ordinates.

48

Developing Numeracy
Measures, Shape and Space
Year 5
© A & C Black

Three in a line game

• **Play this game with a partner.**

☆ Take turns to throw two dice. Choose which number is the **horizontal** co-ordinate and which is the **vertical** co-ordinate.

☆ Write your co-ordinates on the chart. Mark the point on the grid with a cross in your colour.

☆ The winner is the first player to get three in a line.

You need:
two dice
two coloured pencils

Player 1	Player 2

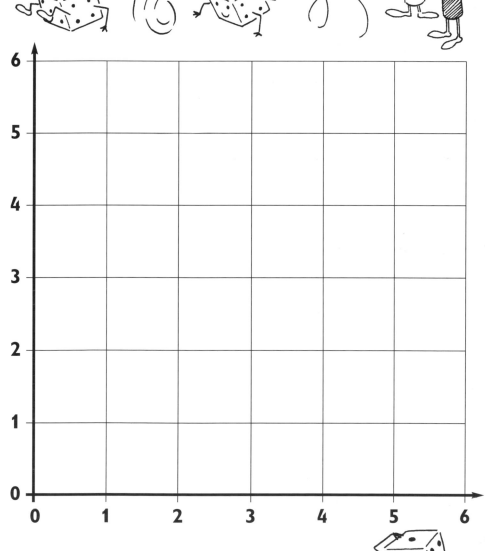

• **How many goes did it take?** ☐

Teachers' note Each pair of children needs one copy of the sheet. Before the activity, discuss with the children how many different points can be thrown with the dice, i.e. 36, and approximately how many throws they think they may need before obtaining three in a line. Explain that both children may mark the same co-ordinate on the grid.

Developing Numeracy
Measures, Shape and Space
Year 5
© **A & C Black**

Polygon points

- **Plot the points.**
- **Write the final point which completes the polygon.**
- **Draw the polygon. Use a ruler.**

On the last grid draw your own polygon. Write its name and the co-ordinates.

square (2, 3) (5, 3) (5, 6) (,   )

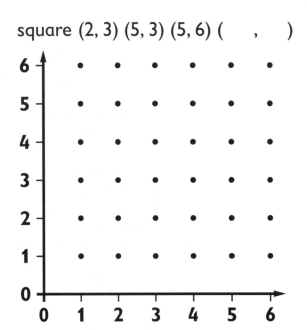

rectangle (6, 2) (5, 1) (3, 5) (,   )

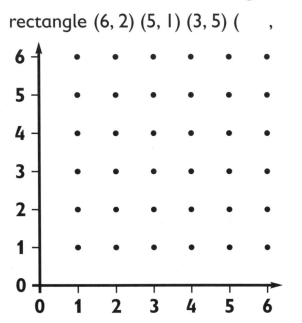

pentagon (1, 4) (3, 6) (1, 2)
(5, 2) (,   )

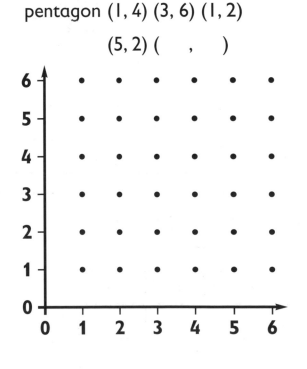

_____

_____

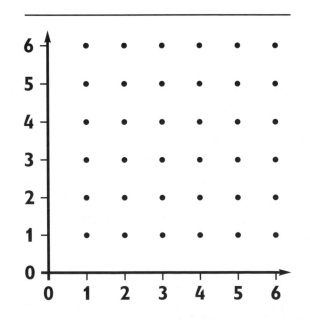

Teachers' note The children could create their own polygons with missing points for a partner to solve. As a further extension, the children could find the area of each polygon. Discuss how many points are needed for particular shapes, for example, hexagons, triangles or trapeziums.

**Developing Numeracy
Measures, Shape and Space
Year 5**
© A & C Black

Dizzy Dog

Dizzy Dog has been programmed
to turn and move forward.

• Write the angle he turns

in right angles and degrees .

1.

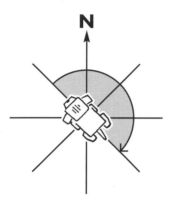

right angles	2
degrees	180°

2.

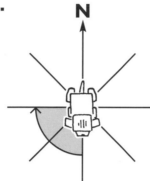

right angles	
degrees	

3.

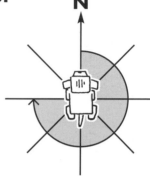

right angles	
degrees	

4.

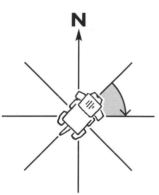

right angles	
degrees	

5.

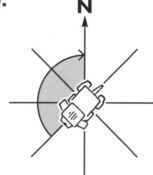

right angles	
degrees	

6.

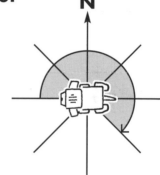

right angles	
degrees	

Now try this!

• **Write Dizzy Dog's start and finish direction
for each move.**

	1	2	3	4	5	6
Start	NW					
Finish	SE					

Teachers' note Before beginning the activity, emphasise the relationship between degrees and right angles, and explain that half a right angle is 45°. Some children may need to be reminded that there are 45° (half of 90°) between each of these points.

**Developing Numeracy
Measures, Shape and Space
Year 5
© A & C Black**

51

The angle game

- **Play this game with a partner.**

You need:
a pencil
a ruler
a protractor.

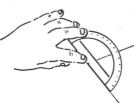

☆ Cut out the cards. Place them in a pile face down.

☆ Take turns to reveal a card. Use a pencil and ruler
   to draw an angle you **estimate** will match the card.

☆ Your partner uses a protractor to
   measure your angle.
   If your angle is correct to within 10°
   you score one point.

☆ The winner is the player with the most points after eight turns each.

130°	85°	60°	55°
25°	115°	140°	80°
70°	95°	40°	135°
105°	150°	75°	110°

Teachers' note The angle cards can be used in different ways, for example, to sort into acute and
obtuse angles. They can be used in games, where players reveal a card and have to say the angle
to add to make 180° (a straight line); or in activities requiring children to find sets of three angles
which could form the angles of a triangle.

Developing Numeracy
Measures, Shape and Space
Year 5
© A & C Black

Alien angles

- **Draw a path from each alien to its space ship.**

Look at the shaded angles.

- **Use a different colour for each alien.**

  They can move vertically, horizontally or diagonally.

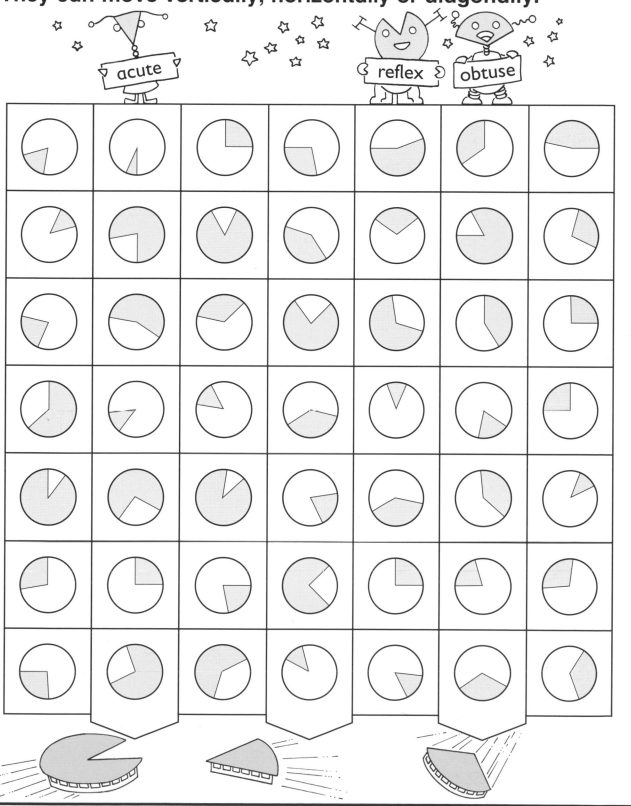

acute   reflex   obtuse

Teachers' note Ensure the children understand that the shaded angles must match the alien's sign. A clock face with moveable hands can be used to introduce the page. Explain the fact that at any time there are two angles between the hands, a larger and a smaller one. Discuss, while rotating the hands, at what times the smaller angle is acute and at what times it is obtuse.

Developing Numeracy
Measures, Shape and Space
Year 5
© A & C Black

Angles on a straight line

- **Calculate the missing angles.**

1.
130° 50°

2.
45°

3.
115°

4.
35°

5.
46°

6.
28°

7.
85° 64°

8.
39° 58°
37°

 Now try this!

- **Draw a straight line. Mark the centre point.**
- **Use a protractor to draw a** ⎡40°⎤ **angle from the centre.**
- **Calculate the other angle. Check with a protractor.**
- **Repeat for angles of** ⎡65°⎤ **and** ⎡125°⎤ **.**

Teachers' note Before beginning the activity, discuss the fact that angles on a straight line total 180°. Once children recognise this, it is a relatively easy step to recognise that the angles at a point total 360°. It can be shown by joining two of the diagrams.

**Developing Numeracy
Measures, Shape and Space
Year 5
© A & C Black**

A pizza cake!

- **Estimate the angle of each slice. Ring your answer.**

- **Use a protractor to measure the angle. Tick the correct angle.**

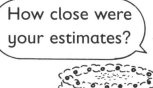
One has been done for you.

1.

85°
(85°) or 95°

2.
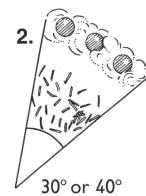
30° or 40°

3.
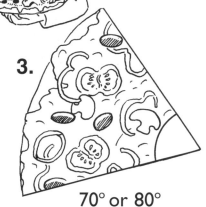
70° or 80°

4.
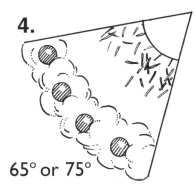
65° or 75°

5.
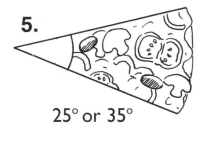
25° or 35°

6.

95° or 105°

7.
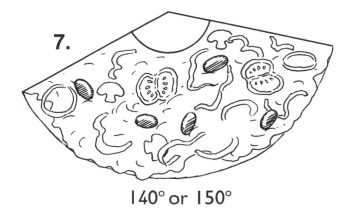
140° or 150°

8.
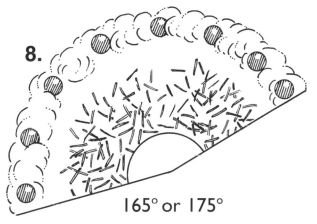
165° or 175°

Now try this!

- **Draw five different angles. Estimate each to the** nearest 5° .
- **Use a protractor to measure each angle to the** nearest 5° .

How close were your estimates?

Teachers' note The correct measures can be demonstrated by producing this page as an OHT, then placing a protractor on top, thus projecting the measurement of the angle onto the screen.

Developing Numeracy
Measures, Shape and Space
Year 5
© A & C Black

Mirror, mirror

- **Draw the** reflection **on the other side of the mirror line.**

One has been done for you.

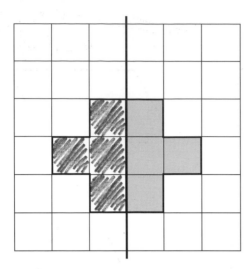

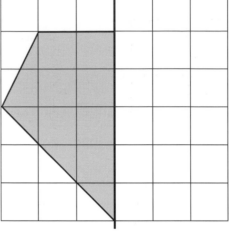

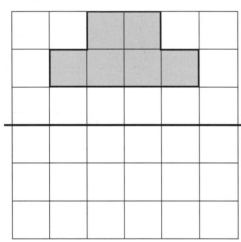

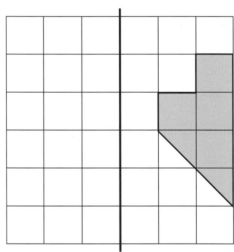

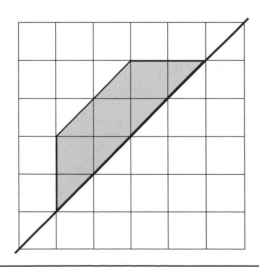

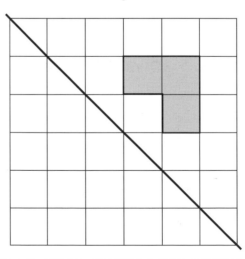

 • **On squared paper, draw three shapes and their reflections.**

Teachers' note The children can use a mirror placed along the mirror line to show, or confirm, the reflection. Remind them that each point is reflected to a point the same distance away from the mirror line on the other side, and that reflected lines will always be the same length.

Developing Numeracy
Measures, Shape and Space
Year 5
© **A & C Black**

Lines of symmetry

• **Draw all the** lines of symmetry **on each polygon.**

One has been done for you.

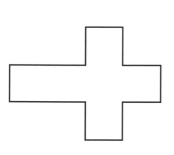
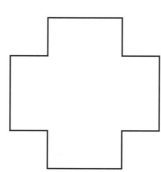
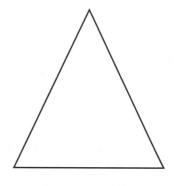

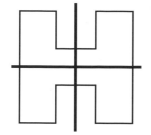

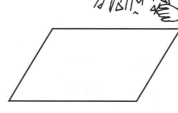

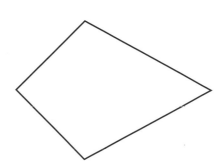
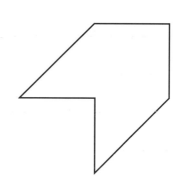

• **Draw polygons with these lines of symmetry.**

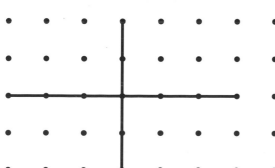
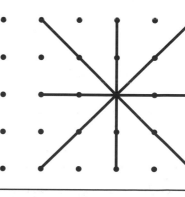

Now try this!

• **Draw a different polygon with four lines of symmetry.**

Teachers' note If in doubt, the shapes can be cut out and folded about the suspected line of symmetry. One side should fold exactly on top of the other side. Note that many children mistakenly believe that a parallelogram has line symmetry.

**Developing Numeracy
Measures, Shape and Space
Year 5**
© A & C Black

57

Symmetrical mosaics

- **Cut out the tiles at the bottom of the page.**
- **Arrange the tiles in squares to make different** `symmetrical` **patterns.**
- **Draw the patterns.**

Teachers' note Discuss which of the children's patterns have one line of symmetry, and which have more. As an extension, the children could design their own simple tile and create their own symmetrical patterns.

**Developing Numeracy
Measures, Shape and Space
Year 5
© A & C Black**

Patchwork quilts

- **Complete each pattern so that it is** $\boxed{\text{symmetrical}}$ **about both lines of symmetry.**

One has been done for you.

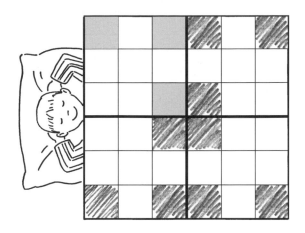

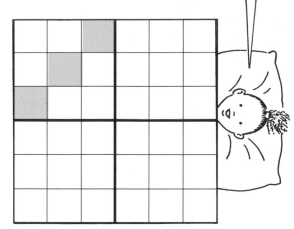

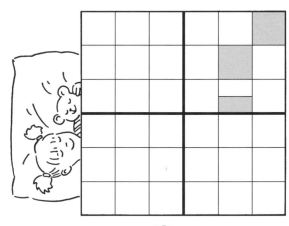

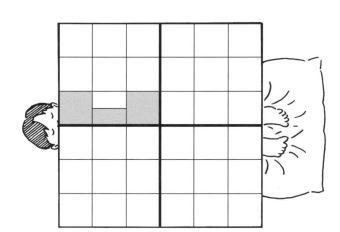

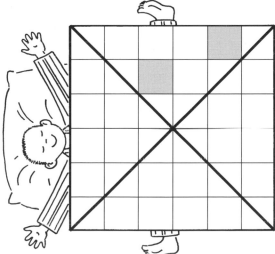

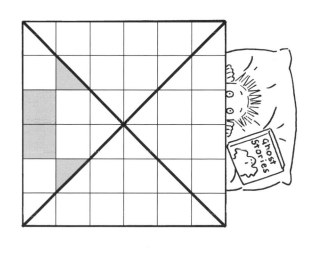

 **Now try this!**

- **On squared paper, create three symmetrical patterns of your own. Each pattern must have two lines of symmetry at right angles.**

Teachers' note Suggest that the children work systematically by drawing the reflection of each shaded square in one line, and then drawing the reflection in the other line. If the children find the last two difficult, suggest that they rotate the page so that the mirror lines are horizontal and vertical.

Developing Numeracy Measures, Shape and Space Year 5 © A & C Black

Symmetrical shapes

- **Join the dots to make shapes which are** $\boxed{\text{symmetrical}}$ **about the line of symmetry.**

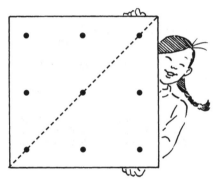
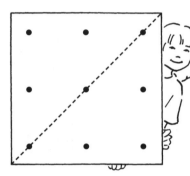
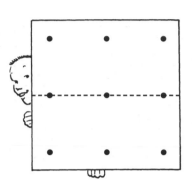
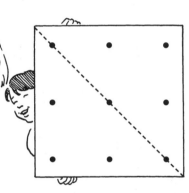

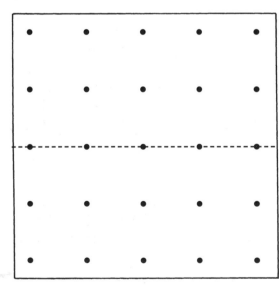

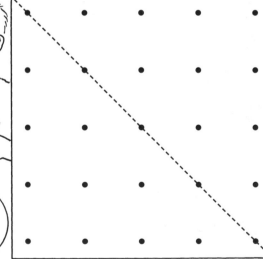

Try some interesting shapes.

Now try this!

- **Find the** $\boxed{\text{area}}$ **of each shape. Write it in units^2.**

Write the area inside the shape.

Teachers' note Suggest to the children that they work in pencil first. Alternatively, they could experiment by creating shapes on a geoboard before drawing them on the grids.

**Developing Numeracy
Measures, Shape and Space
Year 5
© A & C Black**

Answers

p 6

Actual lengths are as follows:

a = 34 mm	**b** = 36 mm	**c** = 25 mm
d = 42 mm	**e** = 10 mm	**f** = 20 mm
g = 24 mm	**h** = 61 mm	**i** = 11 mm
j = 27 mm	**k** = 56 mm	**l** = 130 mm

p 7

Jumps in centimetres:

Jump 1	1 m	100 cm
Jump 2	$\frac{1}{2}$ m	50 cm
Jump 3	$2\frac{1}{2}$ m	250 cm
Jump 4	2·3 m	230 cm

Jump 5	0·5 m	50 cm
Jump 6	$1\frac{1}{4}$ m	125 cm
Jump 7	$\frac{7}{10}$ m	70 cm
Jump 8	4 m	400 cm

Jumps in millimetres:

Jump 9	3 m	3000 mm
Jump 10	25 cm	250 mm
Jump 11	$4\frac{3}{4}$ m	4750 mm
Jump 12	2·6 m	2600 mm

Jump 13	$1\frac{1}{2}$ m	1500 mm
Jump 14	0·9 cm	9 mm
Jump 15	8 cm	80 mm
Jump 16	$4\frac{1}{2}$ cm	45 mm

Now try this!

Centimetre jumps:

Jump 1 – 1 m	Jump 2 – 1 m	Jump 3 – 3 m	Jump 4 – 2 m
Jump 5 – 1 m	Jump 6 – 1 m	Jump 7 – 1 m	Jump 8 – 4 m

Millimetre jumps:

Jump 9 – 3 m	Jump 10 – 0 m	Jump 11 – 5 m	Jump 12 – 3 m
Jump 13 – 2 m	Jump 14 – 0 m	Jump 15 – 0 m	Jump 16 – 0 m

p 8

1. 200 m	**2.** 100 m	**3.** 180 m	**4.** 750 m
5. 325 m	**6.** 170 m	**7.** 2	**8.** 20
9. 13	**10.** 10	**11.** 25	**12.** 100
13. 75	**14.** $4\frac{1}{2}$		

Now try this!

8 lengths a day = 2800 m or 2·8 km
11 widths a day = 1540 m or 1·54 km

p 9

1. 1000 m	**2.** 5300 m
3. 2800 m	**4.** 3050 m
5. 4190 m	**6.** 2005 m

Tiptree: 3000 m
Tenderfoot: 7200 m
Hopeville: 3250 m
Blackcap: 5250 m
Denfield: 4500 m
Tiffield: 4300 m
Claw-on-Sea: 5000 m
Hopton: 1750 m

Now try this!

Tiptree: 6000 m / 6 km
Tenderfoot: 14 400 m / 14·4 km
Hopeville: 6500 m / 6·5 km
Blackcap: 10 500 m / 10·5 km
Denfield: 9000 m / 9 km
Tiffield 8600 m / 8·6 km
Claw-on-Sea: 10 000 m / 10 km
Hopton: 3500 m / 3·5 km

p 11

Cheese:

1 kg = 1000 g	$2\frac{1}{2}$ kg = 2500 g	10 kg = 10 000 g
$\frac{3}{4}$ kg = 750 g	$6\frac{1}{4}$ kg = 6250 g	5 kg 300 g = 5300 g
4·7 kg = 4700 g	$3\frac{1}{4}$ kg = 3250 g	

Mice:

$2\frac{1}{2}$ kg and 1 kg = 3500 g	10 kg and 4·7 kg = 14 700 g
$\frac{3}{4}$ kg and $6\frac{1}{4}$ kg = 7000 g	$3\frac{1}{4}$ kg and 5 kg 300 g = 8550 g

p 14

800 ml	600 ml	250 ml
350 ml	140 ml	80 ml

Now try this!

2940 ml = 2·94 l

p 15

1. 1000 ml	**2.** 2500 ml	**3.** 700 ml
4. 2250 ml	**5.** 4300 ml	**6.** 900 ml
7. 1300 ml	**8.** 1050 ml	**9.** 3400 ml
10. 4600 ml	**11.** 3000 ml	**12.** 4500 ml

p 16

1 pint is approximately 500 ml

	Pints	Litres
Baby Quaff	2 pints	1 litre
Mrs Quaff	4 pints	2 litres
Mr Quaff	9 pints	$4\frac{1}{2}$ litres
Quentin Quaff	3 pints	$1\frac{1}{2}$ litre
Growler the dog	$\frac{1}{2}$ pint	$\frac{1}{4}$ litre

p 17

1. 6:15 pm	**2.** 11:45 am	**3.** 7:25 am
4. 8:25 pm	**5.** 1:20 pm	**6.** 11:50 am

Now try this!

Kooky Bear: 08:25	Dinosaurs: 12:15
Bozz and Buzz: 18:45	Pet Choice: 14:05
Sport for Kids: 12:05	Comedy Half-hour: 20:55

p 19

1. 3 hrs 27 mins	**2.** 3 hrs 30 mins	**3.** 0 hrs 16 mins
4. 0 hrs 26 mins	**5.** 0 hrs 35 mins	**6.** 0 hrs 49 mins
7. 0 hrs 33 mins	**8.** 1 hr 06 mins	**9.** 0 hrs 31 mins
10. 1 hr 35 mins		

Now try this!

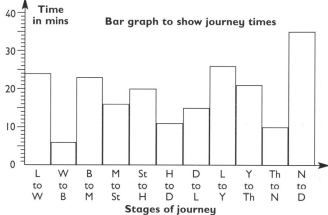

Bar graph to show journey times

p 20

1. 60 secs	**2.** 300 secs	**3.** 30 secs	**4.** 3600 secs
5. 60 mins	**6.** 210 mins	**7.** 135 mins	**8.** 1440 mins
9. 24 hrs	**10.** 168 hrs	**11.** 48 hrs	**12.** 12 hrs
13. 7 days	**14.** 31 days	**15.** 30 days	**16.** 90 or 91 days

p 21

1. 48 m	**2.** 84 m	**3.** 46 m
4. 70 m	**5.** 56 m	**6.** 48 m

Now try this!

1. £144	**2.** £252	**3.** £138
4. £210	**5.** £168	**6.** £144

p 22

a 160 mm	**b** 130 mm	**c** 120 mm
d 130 mm	**e** 180 mm	**f** 160 mm

Correct order: c; b and d; a and f; e

61

p 23

Shape	Area	Perimeter	Number of sides	Name
a	10 cm²	14 cm	4 sides	rectangle
b	9 cm²	12 cm	4 sides	square
c	6 cm²	12 cm	8 sides	octagon
d	5 cm²	12 cm	12 sides	dodecagon
e	7 cm²	12 cm	6 sides	hexagon
f	8 cm²	14 cm	8 sides	octagon
g	10 cm²	16 cm	8 sides	octagon
h	15 cm²	16 cm	4 sides	rectangle
i	5 cm²	10 cm	6 sides	hexagon

Now try this!
The perimeters of the shapes will not all be the same.

p 24
Actual areas are as follows:

a 6 cm² **b** 4 cm² **c** 9 cm² **d** 25 cm²
e 10 cm² **f** 20 cm² **g** 24 cm² **h** 28 cm²

Now try this!
Area of large rectangle is 126 cm².

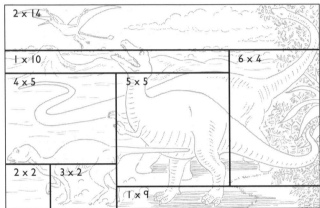

p 25
1. 9 m **2.** 9 m **3.** 8 m **4.** 8 m
5. 9 m **6.** 15 m **7.** 20 m **8.** 12 m

p 26

Perimeter	Area
8 cm	4 cm²
20 cm	25 cm²
80 cm	400 cm²
40 cm	100 cm²
4 cm	1 cm²
44 cm	121 cm²

Area	Perimeter
9 cm²	12 cm
64 cm²	32 cm
36 cm²	24 cm
144 cm²	48 cm
81 cm²	36 cm
900 cm²	120 cm

Now try this!
1 x 24 has a perimeter of 50 cm; 2 x 12 has a perimeter of 28 cm;
3 x 8 has a perimeter of 22 cm; 4 x 6 has a perimeter of 20 cm.
The children may also make rectangles using half centimetres.

p 27
1. hexagon **2.** triangle **3.** pentagon
4. octagon **5.** quadrilateral **6.** decagon

p 28
1. rectangle **2.** right-angled triangle **3.** pentagon
4. quadrilateral **5.** square **6.** octagon

p 29
Check polygons are drawn accurately. Shape names are as follows:
3 sides: triangle   4 sides: quadrilateral (may specify type)
5 sides: pentagon   6 sides: hexagon   7 sides: heptagon
8 sides: octagon   9 sides: nonagon   10 sides: decagon

Now try this!
The maximum number of sides is 15.

p 31

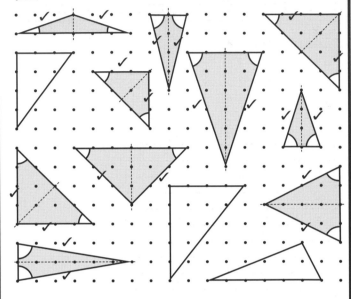

p 32
The right-angled triangles are: 1, 2, 3, 5, 6, 7, 9
Check the right-angled triangles that the children have drawn.

p 33

	Number of angles the same	Number of right angles	Number of sides the same	Is it symmetrical?	Name
a	3	0	3	yes	equilateral
b	2	0	2	yes	isosceles
c	0	0	0	no	scalene
d	0	1	0	no	right-angled / scalene
e	0	1	0	no	right-angled / scalene
f	3	0	3	yes	equilateral
g	2	0	2	yes	isosceles
h	0	1	0	no	right-angled / scalene
i	0	0	0	no	scalene

p 34
There are many possibilities. Check children have shaded different
types of triangles and have named them correctly.

Number of straws	Sides of triangles
3	(1, 1, 1) equilateral
4	impossible
5	(1, 2, 2) isosceles
6	(2, 2, 2) equilateral
7	(1, 3, 3) isosceles, (2, 2, 3) isosceles
8	(2, 3, 3) isosceles
9	(1, 4, 4) isosceles, (2, 3, 4) scalene, (3, 3, 3) equilateral
10	(2, 4, 4) isosceles, (3, 3, 4) isosceles
11	(2, 4, 5) scalene, (3, 4, 4) isosceles, (3, 3, 5) isosceles, (5, 5, 1) isosceles
12	(2, 5, 5) isosceles, (3, 4, 5) scalene, (4, 4, 4) equilateral

p 36
There are many possibilities.

p 38
1. a **2.** f **3.** n **4.** t **5.** j
6. e **7.** r **8.** p **9.** c **10.** l

p 39

	Parallel lines	Perpendicular lines
A		
B		
C		
D		
E	✔	✔
F	✔	✔
G		
H	✔	✔
I	✔	✔
J		
K		
L		✔
M	✔	
N	✔	
O		
P		
Q		
R		
S		
T		✔
U		
V		
W	✔	
X		
Y		
Z	✔	

p 40
There are many possibilities.

p 41

Sides	3	4	5	6	7	8
Diagonals	0	2	5	9	14	20

Now try this!
nonagon: 27 diagonals
decagon: 35 diagonals

p 42

		Length of rectangle					
		1	2	3	4	5	6
Width of rectangle	1	1	2	3	4	5	6
	2	2	2	4	4	6	6
	3	3	4	3	6	7	6
	4	4	4	6	4	8	8
	5	5	6	7	8	5	10
	6	6	6	6	8	10	6

Now try this!
Patterns are as follows:
• symmetry about the diagonal from top left to bottom right;
• n x n rectangle passes through n squares, e.g. 4 x 4 passes through 4 squares.

p 43
1. prism (triangular) **2.** cube*
3. cylinder* **4.** pyramid (triangular-based)
5. cone **6.** prism (pentagonal)
7. cuboid* **8.** cylinder*
9. pyramid (pentagonal-based) **10.** pyramid (square-based)
11. cone **12.** prism (hexagonal)
* These can also be prisms.

p 44
a, d, e, f, g, i, j and l make open cubes.
c makes a closed cube.

p 45

	a	b	c	d	e	f	g	h	i
Number of cubes	4	6	5	6	6	7	7	11	6
Number of cubes to make a cuboid	4	3	7	6	6	5	5	1	3

p 46
Properties of a tetrahedron are:
• 4 faces
• 4 vertices
• 6 edges
• each face is an equilateral triangle

p 48
1. (3, 2) **2.** (3, 5) **3.** (8, 3) **4.** (1, 1)
5. (2, 6) **6.** (6, 7) **7.** (10, 9) **8.** (6, 4)
9. (2, 9) **10.** (8, 5) **11.** (5, 10) **12.** (9, 7)

p 50
square: (2, 6)
rectangle: (2, 4)
pentagon: lots of possibilities, e.g. (4, 5)
Check children's own polygon and co-ordinates for final grid.

p 51
1. 2, 180° **2.** 1, 90° **3.** 3, 270°
4. $\frac{1}{2}$, 45° **5.** 1$\frac{1}{2}$, 135° **6.** 2$\frac{1}{2}$, 225°
Now try this!

	1	2	3	4	5	6
Start	NW	S	N	NE	SW	W
Finish	SE	W	W	E	N	SE

p 54
1. 50° **2.** 135° **3.** 65°
4. 55°, 90° **5.** 44°, 90°
6. 62°, 90° **7.** 31° **8.** 46°
Now try this!
The other angles are: 140°, 115°, 55°

p 55
1. 85° **2.** 40° **3.** 70°
4. 65° **5.** 35° **6.** 105°
7. 140° **8.** 165°

Lines of symmetry are as follows:

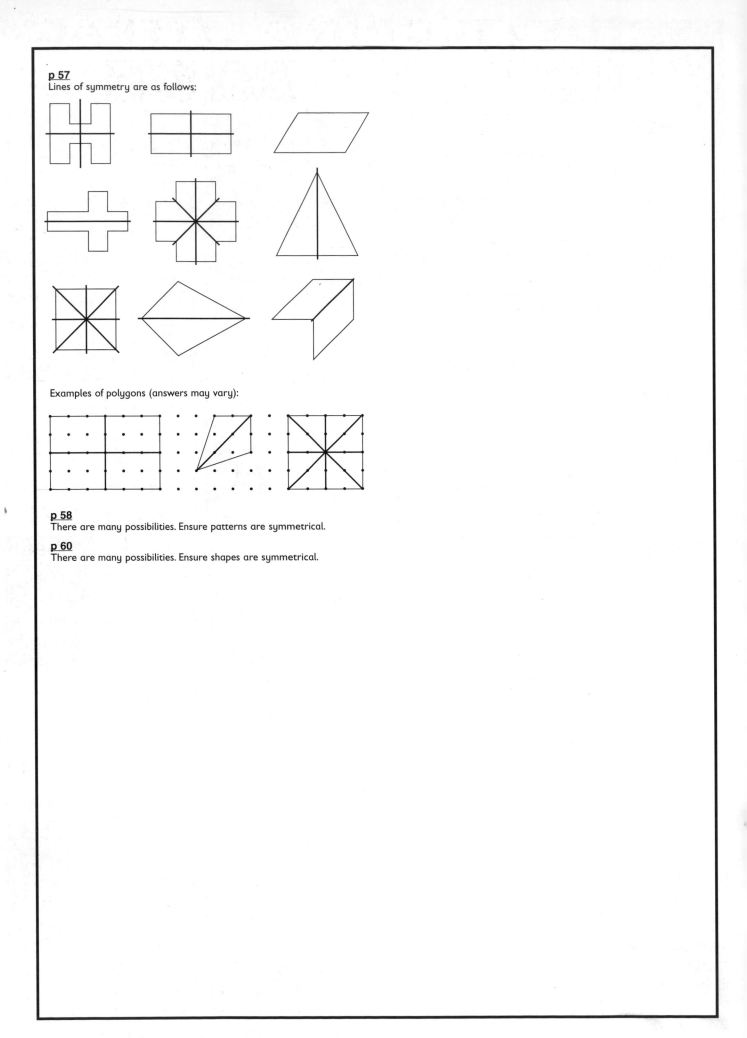

Examples of polygons (answers may vary):

p 58
There are many possibilities. Ensure patterns are symmetrical.

p 60
There are many possibilities. Ensure shapes are symmetrical.